TAXICAB GEOMETRY

An Adventure in Non-Euclidean Geometry

EUGENE F. KRAUSE

Dover Publications, Inc., New York

This Dover edition, first published in 1986, is an unabridged and corrected republication of *Taxicab Geometry*, published by Addison-Wesley Publishing Company, Menlo Park, California, in 1975.

Manufactured in the United States of America
Dover Publications, Inc., 31 East 2nd Street, Mineola, N.Y. 11501

Library of Congress Cataloging-in-Publication Data

Krause, Eugene F., 1937–
    Taxicab Geometry.

    Includes index.
    Summary: Develops a simple non-Euclidean geometry and explores some of its practical applications through graphs, research problems, and exercises. Includes selected answers.
    1. Geometry, Non-Euclidean—Juvenile literature. [1. Geometry, Non-Euclidean]   I. Title.
QA685.K7   1986   516.9   86-13480
ISBN 0-486-25202-7

about this book

This book has a triple purpose: to develop a very simple, concrete non-Euclidean geometry; to explore a few of its many real-world applications; to pose some of the original, yet accessible research questions that abound in this new geometry. The only prerequisite is some familiarity with Euclidean geometry.

about the author

Eugene F. Krause is Professor of Mathematics at the University of Michigan, Ann Arbor, Michigan.

materials

The exercises in this book require graph paper, a ruler, a compass, and a protractor.

TO THE TEACHER

TO FULLY appreciate Euclidean geometry one needs to have some contact with a non-Euclidean geometry. Ideally the non-Euclidean geometry chosen should (1) be very close to Euclidean geometry in its axiomatic structure, (2) have significant applications, and (3) be understandable by anyone who has gone through a beginning course in Euclidean geometry. Condition (1) rules out the various finite geometries as well as the (elliptic) geometry of the sphere.

The other well-known non-Euclidean geometry, hyperbolic geometry, meets condition (1), differing from Euclidean geometry only in its formulation of the parallel postulate, and condition (2), having applications in physics and astronomy. Besides these virtues, its emergence in the 1820's marked a historic step in the evolution of mathematical thought. Unfortunately, hyperbolic geometry fails to meet criterion (3). The concrete embodiment of hyperbolic geometry in the Poincaré model requires much more than just a knowledge of Euclidean geometry in order to be understood. Both the theory and the applications of hyperbolic geometry are quite sophisticated.

Taxicab geometry, on the other hand, is a non-Euclidean geometry that meets all three conditions very nicely. First, it too differs from Euclidean geometry in just one axiom—in this case the "side-angle-side" axiom. Second, it has a wide range of applications to problems in urban geography. While Euclidean geometry appears to be a good model of the "natural" world, taxicab geometry is a better model of the artificial urban world that man has built. Third, taxicab geometry is easy to understand. There are no prerequisites beyond a familiarity with Euclidean geometry and an acquaintance with the coordinate plane. This accessibility of taxicab geometry to high-school students, together with its novelty, makes it a rich source of original research problems which are within a student's grasp.

Since taxicab geometry is so nicely suited on all three grounds, it is puzzling that it has not yet been systematically developed and disseminated. The "taxicab metric," of course, is well known to every student of introductory topology, as are a few of its most elementary properties. In fact, a whole family of "metrics," which includes the taxicab metric, was published by H. Minkowski (1864–1909). But apparently no one has yet set up a full geometry based on the taxicab metric. It would seem that the time has come to do so.

In order to give creativity and originality a chance, this booklet consists mostly of exercises and questions; there is little formal exposition. To work through this material is to participate in the development of taxicab geometry. I wish your students good luck with their mathematical research!

Ann Arbor, Michigan   E.F.K.
April 1986

THE USUAL way to describe a (plane) geometry is to tell what its *points* are, what its *lines* are, how *distance* is measured, and how *angle measure* is determined. When you studied Euclidean coordinate geometry the points were the points of a coordinatized plane. Each of these points could be designated either by a capital letter or by an ordered pair of real numbers (the "coordinates" of the point). For example, in Fig. 1, $P = (-2, -1)$ and $Q = (1, 3)$. The lines were the usual long, straight, skinny sets of points; angles were measured in degrees with a (perfect) protractor; and distances either were measured "as the crow flies" with a (perfect) ruler or were calculated by means of the Pythagorean Theorem.

For example, in Fig. 1 the distance from P to Q could be found by considering a right triangle having $\overline{PQ}$ as its hypotenuse. The dotted segments are the legs of one such triangle. (Are there any other such right triangles?) These legs clearly have lengths 3 and 4. Thus, by the Pythagorean Theorem, the Euclidean distance from P to Q is $\sqrt{3^2 + 4^2} = 5$. We shall use the symbol d_E to represent the Euclidean distance function. Thus, in our example we would write

$$d_E(P, Q) = 5,$$

and read it "The Euclidean distance from P to Q is 5."

Taxicab geometry is very nearly the same as Euclidean coordinate geometry. The points are the same, the lines are the same, and angles are measured in the same way. Only the distance function is different. In Fig. 1 the taxicab distance from P to Q, written $d_T(P, Q)$, is determined not as the crow flies, but instead as a taxicab would drive. We count how many blocks it would have to travel horizontally and vertically to get from P to Q. The dotted segments suggest one taxi route. Clearly

$$d_T(P, Q) = 7.$$

"The taxi distance from P to Q is 7."

1
WHAT IS TAXICAB GEOMETRY?

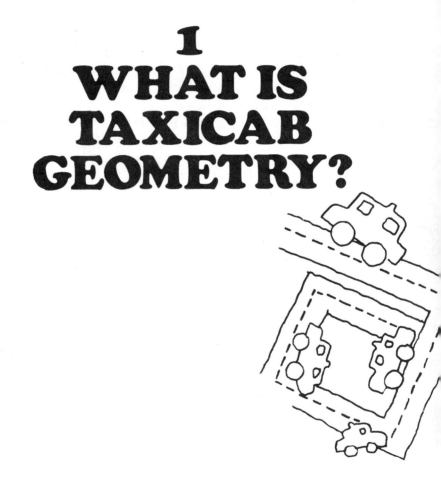

CONTENTS

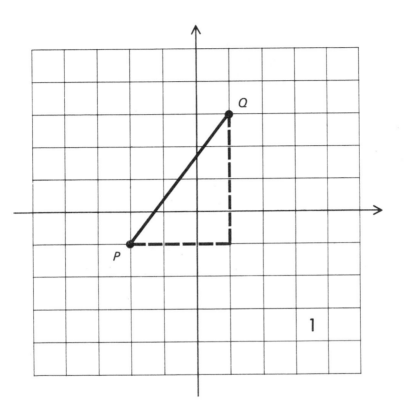

Figure 2 is a reminder that most of the points of the coordinate plane do not have two integer coordinates. In the figure a pair of "arbitrary" points $A = (a_1, a_2)$ and $B = (b_1, b_2)$ is given. What are the coordinates of the point C? Write an expression for the length of $\overline{AC}$ in terms of the coordinates of A and C. Write an expression for the length of $\overline{BC}$ in terms of the coordinates of B and C. The following precise, algebraic definitions of d_T and d_E should now seem reasonable:

(1) $$d_T(A, B) = |a_1 - b_1| + |a_2 - b_2|;$$

(2) $$d_E(A, B) = \sqrt{(a_1 - b_1)^2 + (a_2 - b_2)^2}.$$

We will make use of these careful definitions only very rarely. The reason for inserting them here is to assure ourselves that (1) there is a mathematically respectable foundation underlying taxicab geometry, and (2) there is a definite taxicab distance between any two points, whether they are located at a "street corner" or not.

what is taxicab geometry?

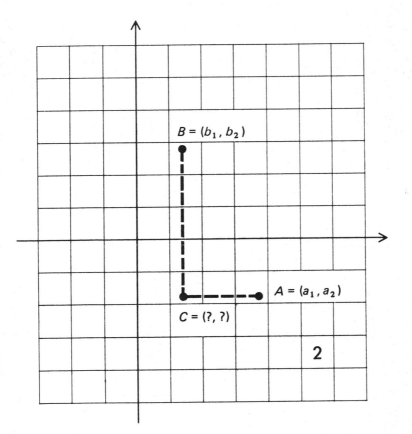

$B = (b_1, b_2)$

$A = (a_1, a_2)$

$C = (?, ?)$

2

exercises

1. On a sheet of graph paper, mark each pair of points P and Q and then find both $d_T(P, Q)$ and $d_E(P, Q)$.

   a) $P = (5, 4)$, $Q = (1, 2)$

   b) $P = (-4, 3)$, $Q = (3, 2)$

   c) $P = (-5, -4)$, $Q = (1, -2)$

   d) $P = (3, -1)$, $Q = (-2, 4)$

   e) $P = (4, -3)$, $Q = (-2, -3)$

2. a) If $d_T(A, B) = d_T(C, D)$ must $d_E(A, B) = d_E(C, D)$?

   b) If $d_E(A, B) = d_E(C, D)$ must $d_T(A, B) = d_T(C, D)$?

   c) Under what conditions on A and B does $d_T(A, B) = d_E(A, B)$?

   d) Is it always true that $d_E(A, B) \leqslant d_T(A, B)$? Try to prove your answer using the formal definitions (1) and (2).

3. For this exercise $A = (-2, -1)$. Mark A on a sheet of graph paper. For each point P below calculate $d_T(P, A)$ and mark P on the graph paper.

   a) $P = (1, -1)$

   b) $P = (-2, -4)$

   c) $P = (-1, -3)$

   d) $P = (0, -2)$

   e) $P = (\frac{1}{2}, -1\frac{1}{2})$

   f) $P = (-1\frac{1}{2}, -3\frac{1}{2})$

   g) $P = (0, 0)$

   h) $P = (-2, 2)$

exercises

4. a) Find some more points at taxi distance 3 from A.

 b) Graph the set of *all* points P at taxi distance 3 from A; that is, graph $\{P|d_T(P, A) = 3\}$. The set notation

$$\text{``}\{P|d_T(P, A) = 3\}\text{''}$$

 is usually read: "the set of all points P such that the taxi distance from P to A is 3."

 c) Graph the set of all points P at Euclidean distance 3 from A; that is, graph $\{P|d_E(P, A) = 3\}$.

 d) Invent a reasonable name for $\{P|d_T(P, A) = 3\}$.

 e) In taxicab geometry, what is a reasonable numerical value for π?

5. Given $A = (-2, -1)$ and $B = (3, 2)$. Graph the following sets of points.

 a) The taxi circle with center A and radius 2.

 b) $\{P|d_T(P, A) = 1\}$

 c) The set of all points P at taxi distance $1\frac{1}{2}$ from A.

 d) The taxi circle with center B and radius 4.

 e) $\{P|d_T(P, B) = 2\frac{1}{2}\}$

6. Given $A = (-2, -1)$ and $B = (3, 2)$.

 a) Calculate $d_T(A, B)$.

 Now on a single sheet of graph paper:

 b) Graph $\{P|d_T(P, A) = 3$ and $d_T(P, B) = 5\}$.

 c) Graph $\{P|d_T(P, A) = 1$ and $d_T(P, B) = 7\}$.

 d) Graph $\{P|d_T(P, A) = 0$ and $d_T(P, B) = 8\}$.

7

e) Graph $\{P|d_T(P, A) = 1\frac{1}{2}$ and $d_T(P, B) = 6\frac{1}{2}\}$.

f) Graph $\{P|d_T(P, A) = 4$ and $d_T(P, B) = 4\}$.

g) Graph $\{P|d_T(P, A) = 5$ and $d_T(P, B) = 3\}$.

h) Graph $\{P|d_T(P, A) + d_T(P, B) = d_T(A, B)\}$.

7. Given $A = (-7, -3)$ and $B = (5, 2)$.

   a) Calculate $d_E(A, B)$.

   b) Graph $\{P|d_E(P, A) + d_E(P, B) = d_E(A, B)\}$.

8. On a sheet of graph paper mark each pair of points A and B and then graph $\{P|d_T(P, A) + d_T(P, B) = d_T(A, B)\}$

   a) $A = (-2, 3)$ and $B = (1, -4)$

   b) $A = (1, -3)$ and $B = (4, 0)$

   c) $A = (2, 1)$ and $B = (6, 1)$

   d) $A = (1, 1)$ and $B = (1, 4)$

9. Given $A = (-2, -1)$ and $B = (3, 2)$

   a) Graph $\{P|d_T(P, A) = 5$ and $d_T(P, B) = 5\}$.

   b) Graph $\{P|d_T(P, A) = 7$ and $d_T(P, B) = 7\}$.

   c) Graph $\{P|d_T(P, A) = 4$ and $d_T(P, B) = 4\}$.

   d) Graph $\{P|d_T(P, A) = d_T(P, B)\}$.

10. Repeat Exercise 9 using d_E in place of d_T. (A compass will be helpful.)

11. For each pair of points A and B, graph $\{P|d_T(P, A) = d_T(P, B)\}$

    a) $A = (0, 0)$ and $B = (4, 2)$

    b) $A = (0, 0)$ and $B = (2, 4)$

exercises

11. (*cont'd.*)

  c) $A = (0, 0)$ and $B = (3, 3)$ (Watch out!)

  d) $A = (-1, 1)$ and $B = (4, 1)$

12. Plot $A = (-3, 0)$ and $B = (1, 2)$ and then graph

$$\{P | d_T(P, A) = 2 \cdot d_T(P, B)\}.$$

13. Repeat Exercise 12 using d_E in place of d_T.

2
SOME
APPLICATIONS

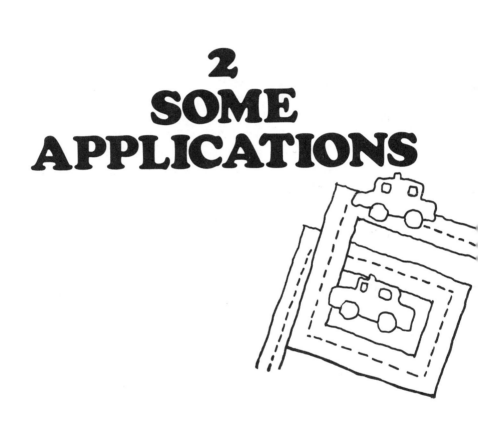

TAXICAB GEOMETRY is a more useful model of urban geography than is Euclidean geometry. Only a pigeon would benefit from the knowledge that the Euclidean distance from the Post Office to the Museum (Fig. 3) is $\sqrt{8}$ blocks while the Euclidean distance from the Post Office to the City Hall is $\sqrt{9} = 3$ blocks. This information is worse than useless for a person who is constrained to travel along streets or sidewalks. For people, taxicab distance is the "real" distance. It is *not* true, for people, that the Museum is "closer" to the Post Office than the City Hall is. In fact, just the opposite is true. (What are the two taxicab distances?)

While taxicab geometry is a better mathematical model of urban geography than is Euclidean geometry, it is not perfect. Many simplifying assumptions have been made about the city. All the streets are assumed to run straight north and south or straight east and west; streets are assumed to have no width; buildings are assumed to be of point size . . . You should not be greatly disturbed by these assumptions. True, no city is exactly like the ideal one we have in mind. Still, many parts of many cities are not too different from it. The things we learn about our ideal model will have application in certain real urban situations.

The process of setting up a mathematical model of a real situation nearly always involves making simplifying assumptions. Without them the mathematical problems tend to be too involved and difficult to solve, or even to set up. In Section 6 we shall see some of the mathematical complications that arise when we alter our ideal model to make it more realistic.

some applications

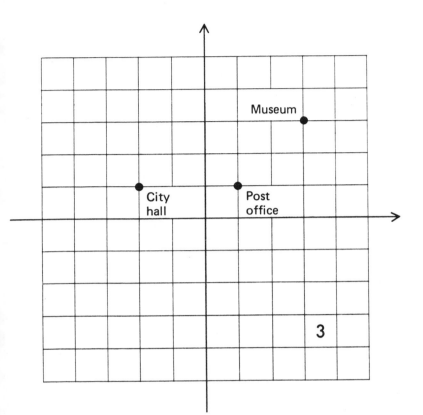

13

exercises

1. List some additional simplifying assumptions that we have made about Ideal City.

2. Alice and Bruno are looking for an apartment in Ideal City. Alice works as an acrobat at amusement park $A = (-3, -1)$. Bruno works as a bread taster in bakery $B = (3, 3)$. (See Fig. 4.) Being ecologically aware, they walk wherever they go. They have decided their apartment should be located so that the distance Alice has to walk to work plus the distance Bruno has to walk to work is as small as possible. Where should they look for an apartment?

3. In a moment of chivalry Bruno decides that the sum of the distances should still be a minimum, but Alice should not have to walk any farther than he does. Now where could they look for an apartment?

4. Alice agrees that the sum of the distances should be a minimum, but she is adamant that they both have exactly the same distance to walk to work. Now where could they live?

5. After a day of fruitless apartment hunting they decide to widen their area of search. The only requirement they keep is that they both be the same distance from their jobs. Now where should they look?

6. After another luckless day they finally agree that all that really matters is that Bruno be closer to his job than Alice is to hers. Now where can they look?

7. The dispatcher for the Ideal City Police Department receives a report of an accident at $X = (-1, 4)$. There are two police cars in the area, car C at $(2, 1)$ and car D at $(-1, -1)$. Which car should she send to the scene of the accident?

14

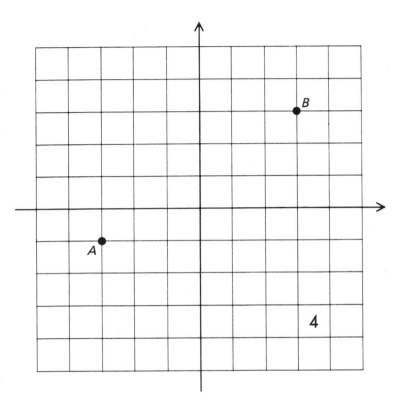

8. A builder wants to put up an apartment building within six blocks of the shopping center $S = (-3, 0)$ and within four blocks of the tennis courts $T = (2, 2)$. Where can he build?

9. The newly elected mayor has promised to install drinking fountains in Ideal City so that every citizen is within three blocks of a drink of water. He discovers that money for civic improvements is rather scarce. His three aides present him with three plans for locating fountains (Figs. 5, 6, and 7). Which should he probably pick, and why?

10. The telephone company wants to set up pay-phone booths so that everyone living within twelve blocks of the center of town is within four blocks of a pay phone. How few booths can they get by with, and where should they be located?

11. A group of students has decided to start a Junior Achievement business of custom finishing furniture. They will buy unfinished furniture at warehouse $W = (-3, 2)$, transport it to their shop S for finishing, and then deliver it to retail store $R = (5, -1)$ for sale. Where should they locate their shop S if they want to minimize the distance they will have to haul furniture?

12. There are three high schools in Ideal City: Fillmore at $(-4, 3)$, Grant at $(2, 1)$, and Harding at $(-1, -6)$. Draw in school-district boundary lines so that each student in Ideal City attends the high school nearest his home.

13. If Burger Baron wants to open a hamburger stand equally distant from each of the three high schools, where should it be located?

14. A fourth high school, Polk High, has just been built at $(2, 5)$. Redraw the school-district boundary lines.

15. (A Plain Geometry problem) Out in the wide open spaces people don't have to walk around buildings. Their distance function is Euclidean distance. Repeat Exercises 12 through 14, assuming that the schools are out on the Great Plains.

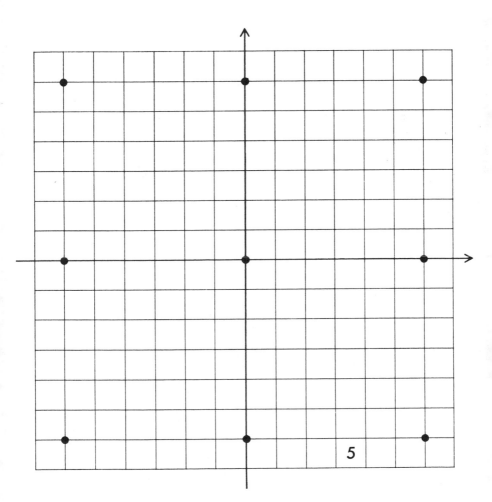

5

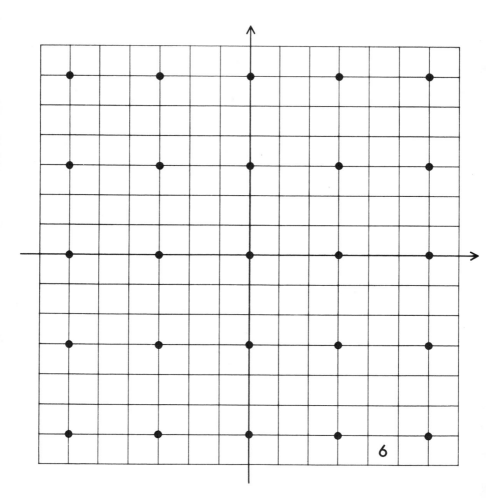

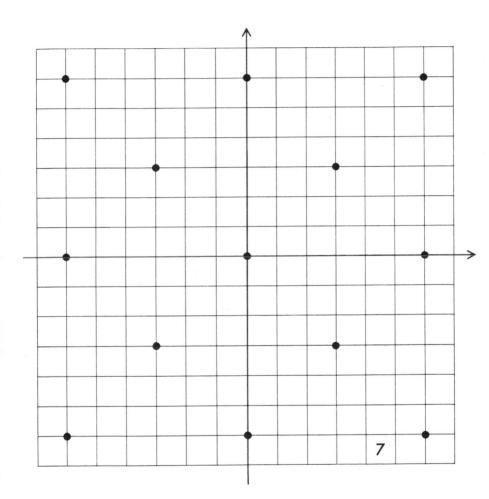

7

3
SOME
GEOMETRIC
FIGURES

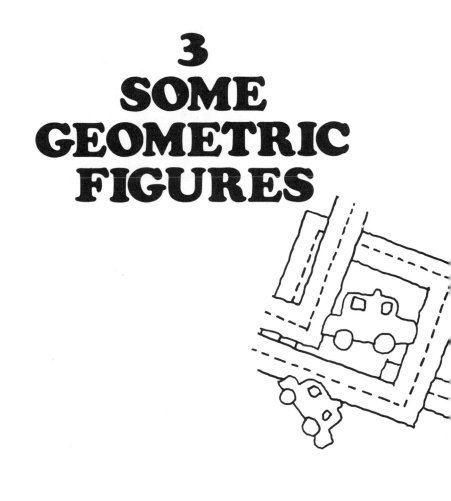

WE HAVE already seen how some familiar geometric figures are transmuted in taxicab geometry. For example, circles in taxicab geometry are squares. As another example, the set of all points equidistant from two given points A and B looks quite different in taxicab geometry than in Euclidean geometry. In Euclidean geometry it is just the perpendicular bisector of $\overline{AB}$. In taxicab geometry it can have a variety of shapes (See Exercise 11 of Section 1), but only rarely turns out to be the perpendicular bisector of $\overline{AB}$.

There are other useful geometric figures which can be defined in terms of distance and which deserve study in both Euclidean and taxicab geometry. One such figure is the *ellipse*. By definition an ellipse is the set of all points the sum of whose distances from two given points is a constant. If we let $A = (-2, -1)$ and $B = (2, 2)$ be the two given points, called the *foci* of the ellipse, then one Euclidean ellipse with foci A and B is

$$\{P \,|\, d_E(P, A) + d_E(P, B) = 6\}.$$

This ellipse is the solid one in Fig. 8. Another Euclidean ellipse with foci A and B is

$$\{P \,|\, d_E(P, A) + d_E(P, B) = 9\}.$$

It is the dotted one in Fig. 8.

A procedure for sketching an ellipse, for example the solid one

$$\{P \,|\, d_E(P, A) + d_E(P, B) = 6\}$$

with foci $A = (-2, -1)$ and $B = (2, 2)$, is as follows.

    i) Use a compass to draw a circle of radius 4 with center A and a circle of radius 2 with center B. (These circles are the solid ones in Fig. 9.) Any point lying on both these circles will be at distance 4 from A and 2 from B; thus the sum of its distances from A and B will be 6, and it will be a point of the ellipse we are after.

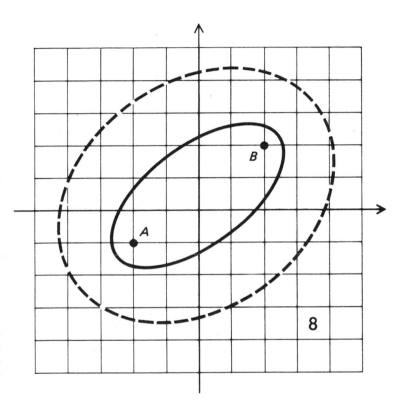

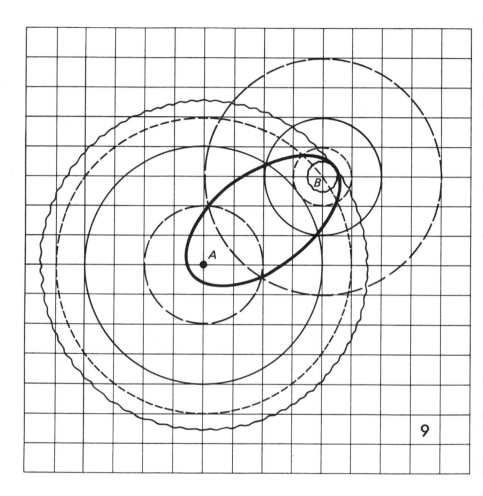

ii) Draw a circle of radius 2 with center A and a circle of radius 4 with center B. (These are the dashed ones in Fig. 9.) The intersection of these two circles contributes two more points to the ellipse.

iii) Draw a circle of radius 5 with center A and a circle of radius 1 with center B. (These are the dotted ones in Fig. 9.) Their intersection contributes two more points to the ellipse.

iv) Draw a circle of radius $5\frac{1}{2}$ with center A and a circle of radius $\frac{1}{2}$ with center B. (These are the wavy ones in Fig. 9.) Note that the intersection of these two circles is a single point.

v) Draw other pairs of circles with centers A and B and radii that add up to 6, to find other points on the ellipse.

vi) When you have found enough points to see the general shape of the ellipse, join the points with a smooth curve.

exercises

1. Mark $A = (-2, -1)$ and $B = (2, 2)$ on a sheet of graph paper. Sketch the ellipse, $\{P | d_E(P, A) + d_E(P, B) = 9\}$, by following these steps.

   a) With a compass, draw lightly a circle with center A and radius $4\frac{1}{2}$, and another circle with center B and radius $4\frac{1}{2}$. Darken their points of intersection.

   b) Repeat (a) using center A, radius 5, and center B, radius 4.

   c) Repeat using center A, radius 6, and center B, radius 3.

   d) Repeat using center A, radius $6\frac{1}{2}$, and center B, radius $2\frac{1}{2}$.

   e) Repeat using center A, radius 7, and center B, radius 2.

   f) Repeat using center A, radius 8, and center B, radius 1.

   g) Sketch the desired ellipse.

2. Using $A = (-2, -1)$ and $B = (2, 2)$ as foci, sketch the following sets on a single sheet of graph paper. Use a different color for each one. (An efficient way to plot all these figures is to begin by drawing a whole family of circles centered at A and having radii $1, 2, \ldots, 8$, and another such family with centers at B.)

   a) $\{P | d_E(P, A) + d_E(P, B) = 11\}$

   b) $\{P | d_E(P, A) + d_E(P, B) = 8\}$

   c) $\{P | d_E(P, A) + d_E(P, B) = 5\frac{1}{2}\}$

   d) $\{P | d_E(P, A) + d_E(P, B) = 5\}$

   e) $\{P | d_E(P, A) + d_E(P, B) = 4\}$

3. Points A and B are the same as in Exercise 2.

   a) Calculate $d_E(A, B)$. Does this calculation shed any light on parts (d), (e), and (c) of Exercise 2?

3. (*cont'd.*)

   b) What do you suppose is the general shape of the ellipse, $\{P | d_E(P, A) + d_E(P, B) = 100\}$?

   c) Kepler's First Law of planetary motion states that the orbit of each planet is an ellipse with the sun at one focus. The Earth's "other" focus is about 5 million kilometers from the sun. The sum of the Earth's distances from its foci is about 300 million kilometers. What is the general shape of the Earth's orbit?

4. Again mark $A = (-2, -1)$ and $B = (2, 2)$ on a sheet of graph paper. Devise a procedure and sketch the *taxicab ellipse*

$$\{P | d_T(P, A) + d_T(P, B) = 9\}.$$

5. On a new sheet of graph paper, again mark $A = (-2, -1)$ and $B = (2, 2)$, and copy the taxicab ellipse of Exercise 4. Now sketch in different colors these other sets.

   a) $\{P | d_T(P, A) + d_T(P, B) = 13\}$

   b) $\{P | d_T(P, A) + d_T(P, B) = 7\}$

   c) $\{P | d_T(P, A) + d_T(P, B) < 13\}$

6. Sketch the taxicab ellipse, $\{P | d_T(P, M) + d_T(P, N) = 10\}$, where $M = (-2, 1)$ and $N = (4, 1)$.

7. While Euclidean ellipses have their applications in the heavens, taxicab ellipses have theirs on Earth. Alice and Bruno still have not found an apartment. Remember that Alice works at $(-3, -1)$ and Bruno works at $(3, 3)$. They have now decided that the sum of the distances that they have to walk to work should be no more than 14 blocks. Where can they look for an apartment?

8. Ajax Industrial Corporation wants to build a factory in Ideal City in a location where the sum of its distances from the rail-

road station $C = (-5, -3)$ and the airport $D = (5, -1)$ is at most 16 blocks. For noise-control purposes, a city ordinance forbids the location of any factory within 3 blocks of the public library $L = (-4, 2)$. Where can Ajax build?

9. On a sheet of graph paper mark $A = (-2, -1)$ and $B = (2, 2)$.

a) Devise a procedure (using your compass) and sketch this figure:

$$\{P | d_E(P, A) - d_E(P, B) = 3\}.$$

b) The figure you just sketched is one branch of a (Euclidean) *hyperbola* with foci A and B. The other branch of this hyperbola is

$$\{P | d_E(P, B) - d_E(P, A) = 3\}.$$

Sketch it.

c) Explain why the set notation

$$\{P | |d_E(P, A) - d_E(P, B)| = 3\}$$

describes the entire hyperbola (both branches).

10. On a new sheet of graph paper, again mark $A = (-2, -1)$ and $B = (2, 2)$ and copy the hyperbola of Exercise 9. Now sketch these sets in different colors.

a) $\{P | |d_E(P, A) - d_E(P, B)| = 1\}$

b) $\{P | |d_E(P, A) - d_E(P, B)| = 0\}$

c) $\{P | |d_E(P, A) - d_E(P, B)| = 4\}$

d) $\{P | |d_E(P, A) - d_E(P, B)| = 5\}$

e) $\{P | |d_E(P, A) - d_E(P, B)| = 6\}$

What is significant about the number 5?

11. Mark $A = (-3, -1)$ and $B = (2, 2)$ on a sheet of graph paper.

Now sketch the *taxicab hyperbola*

$$\{P\,|\,|d_T(P, A) - d_T(P, B)| = 3\}.$$

12. On a new sheet of graph paper again mark $A = (-3, -1)$ and $B = (2, 2)$ and copy the taxicab hyperbola of Exercise 11. Using a different color for each one, sketch the following additional figures.

   a) $\{P\,|\,|d_T(P, A) - d_T(P, B)| = 1\}$

   b) $\{P\,|\,|d_T(P, A) - d_T(P, B)| = 0\}$

   c) $\{P\,|\,|d_T(P, A) - d_T(P, B)| = 2\}$ (Watch out!)

   d) $\{P\,|\,|d_T(P, A) - d_T(P, B)| = 8\}$ (Be careful!)

   e) $\{P\,|\,|d_T(P, A) - d_T(P, B)| = 9\}$

   What is significant about the number 8?

13. Investigate the family of taxicab hyperbolas with foci $A = (-3, 1)$ and $B = (5, 1)$.

14. Investigate the family of taxicab hyperbolas with foci $A = (0, 0)$ and $B = (4, 4)$.

15. Alice and Bruno still don't have an apartment. Their latest agreement is that neither person should have to walk more than 4 blocks farther to work than the other person. Where can they look?

4
DISTANCE FROM A POINT TO A LINE

IN EUCLIDEAN geometry there is a standard method for determining the distance from a point A to a line L. (See Fig. 10.) One first locates the line L' through A perpendicular to L. Then, letting B be the point of intersection of L' and L, one observes that the Euclidean distance from A to L is just the Euclidean distance from A to B. In symbols,

$$d_E(A, L) = d_E(A, B)$$

We shall see in the exercises that the procedure for finding the distance from a point to a line in taxicab geometry is quite different.

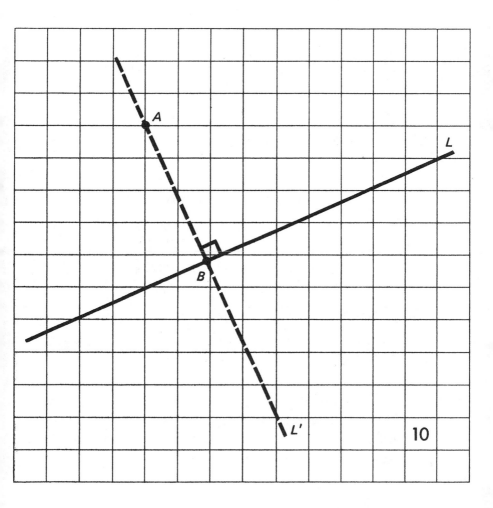

exercises

1. Figure 11 shows a point A and a line L.

   a) Find $d_T(A, B)$.

   b) Find $d_T(A, C)$.

   c) Find $d_T(A, D)$.

   d) Find a point P on L which is "as close as possible" to A in taxicab geometry. (Is P on the perpendicular to L through A?)

   e) What would you say is the taxicab distance from A to L?

2. On a sheet of graph paper, sketch the point $A = (-3, 2)$ and the line L passing through $(-6, -2)$ and $(0, 0)$.

   a) Locate a point P on L that is as close as possible to A in taxicab geometry.

   b) Calculate $d_T(A, L)$.

3. Repeat Exercise 2 for $A = (-3, 2)$ and L the line through $(-2, -1)$ and $(2, 3)$.

4. So far we have considered *procedures* for finding the distance from a point to a line in Euclidean geometry and in taxicab geometry. But as yet we have formulated no *definitions* of what is meant by the distance from a point to a line. Exercises 1 through 3 suggest the following definition for taxicab geometry:

   $d_T(A, L) = $ The smallest of all the $d_T(A, P)$ where $P \in L$.

   This is often abbreviated

   $$d_T(A, L) = \min_{P \in L} d_T(A, P).$$

   State the corresponding definition of $d_E(A, L)$.

5. Another way of thinking about how to find $d_E(A, L)$ is suggested by Fig. 12. Think of slowly inflating a circle with center A until

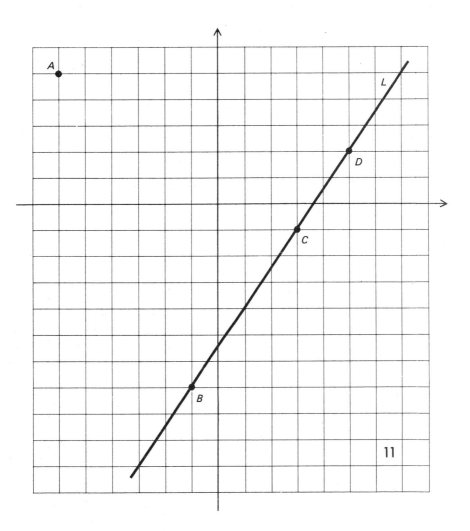

11

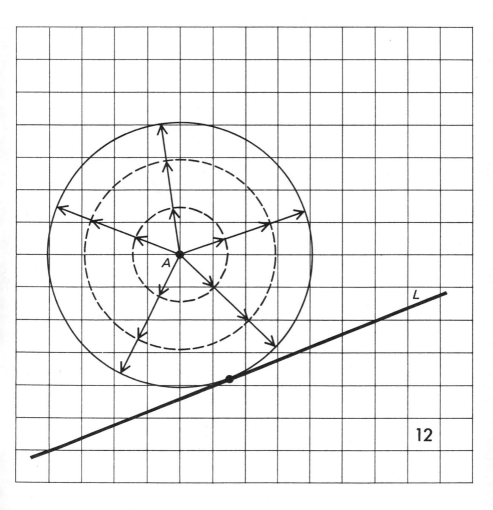

it just touches L. Its radius at that moment is $d_E(A, L)$. Remembering what circles look like in taxicab geometry, reconsider Exercises 1(d), 1(e), 2, and 3 from the point of view of inflating taxicab circles.

6. The Euclidean method for finding $d_E(A, L)$ is often stated loosely as follows. "Measure the distance from A to L along the perpendicular." We would like to formulate a similar loose, easily remembered verbalization for how to find $d_T(A, L)$. To do so, it is convenient to introduce some terminology. For any point P in the coordinate plane, we classify the lines through P under three headings. (See Fig. 13.)

   i) "45° lines"—L_1 and L_2 are the only 45° lines (through P).

   ii) "Steep lines"—Any line (through P) that lies in the shaded region. L_3 is an example of a steep line.

   iii) "Gradual lines" – Any line (through P) that lies in the unshaded region. L_4 is an example of a gradual line.

   Complete these statements that tell how to determine the taxicab distance from a point A to a line L.

   a) "If L is steep, measure the distance from A to L along _____ ."

   b) "If L is gradual, _____ ."

   c) "If L is a 45° line, _____ ."

7. On a sheet of graph paper, draw the line L through $(3, 0)$ and $(1, -4)$. Now,

   a) Sketch $\{P \mid d_T(P, L) = 2\}$

   b) Sketch $\{P \mid d_T(P, L) = 4\}$

   On the same sheet of graph paper, but in a different color,

   c) Sketch $\{P \mid d_E(P, L) = 2\}$

   d) Sketch $\{P \mid d_E(P, L) = 4\}$

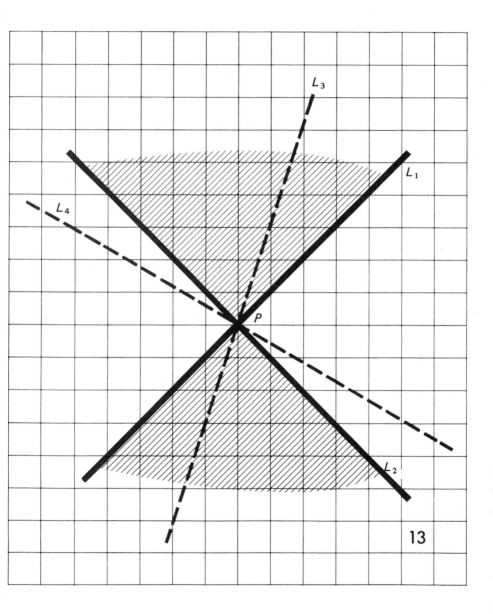

13

8. Repeat Exercise 7 using the line L through $(0, 0)$ and $(3, 1)$.

9. Can you think of a line L for which
$$\{P|d_T(P, L) = 2\} = \{P|d_E(P, L) = 2\}?$$

10. Figure 14 shows a point F and a line L.

   a) Sketch $\{P|d_T(P, F) = 2\}$

   b) Sketch $\{P|d_T(P, L) = 2\}$

   c) Sketch $\{P|d_T(P, F) = 2 \text{ and } d_T(P, L) = 2\}$

   d) Sketch $\{P|d_T(P, F) = d_T(P, L)\}$

11. With F and L as in Exercise 10, sketch
$$\{P|d_E(P, F) = d_E(P, L)\}.$$

(A ruler and compass will be useful.)

12. The figure in Exercise 11 is known as a (Euclidean) *parabola*. F is called its *focus*, L its *directrix*. Thus we shall refer to
$$\{P|d_T(P, F) = d_T(P, L)\}$$
as the *taxi parabola* with focus F and directrix L. Sketch the taxi parabola with the focus F and directrix L given:

   a) $F = (-2, 2)$ L is the line through $(-2, -2)$ and $(2, 2)$;

   b) $F = (0, 4)$ L is the line through $(0, 0)$ and $(2, 0)$.

13. Alice still works as an acrobat at $A = (-3, -1)$, but Bruno has a new job as a conductor on the new mass-transit vehicle which runs along the line L shown in Fig. 15. One of Bruno's fringe benefits is that when he comes to work he can get on the vehicle at the point nearest his home. This sends Alice and Bruno off on another apartment search.

   a) They want to live where the distance Alice has to walk to work plus the distance Bruno has to walk to work is a minimum. Where should they look?

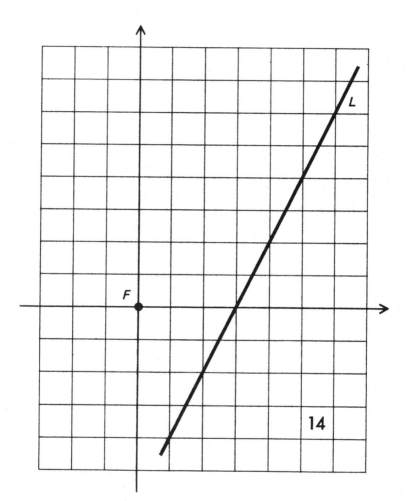

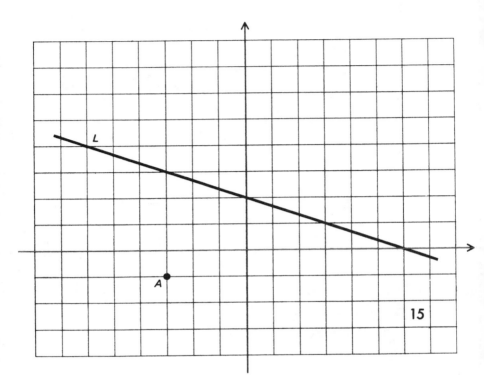

15

13. (*cont'd.*)

b) They change their minds and decide to live where they both have the same distance to walk to work. Where should they look?

c) Where should they look if all that matters is that Alice have a shorter distance to walk than Bruno?

d) Where should they look if they both want to be within 6 blocks of their job?

e) Where should they look if the sum of the distances they have to walk is to be at most 6 blocks?

14. For old times' sake, Alice wants to walk from the amusement park $A = (-3, -1)$ to the bakery $B = (3, 3)$, but she would like to stop along the way to watch the freight train go by on the track shown in Fig. 16. Where should she stop to watch the train if she wants to minimize the distance she walks? How long will her hike from A to B be?

15. Repeat Exercise 14 in the case where the (straight) railroad tracks pass through $(0, -8)$ and $(4, 0)$.

16. In Fig. 17 a point A and a set S are shown. Approximate these two distances.

a) $d_E(A, S)$

b) $d_T(A, S)$

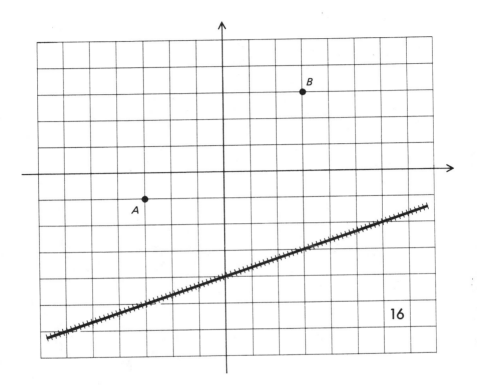

16

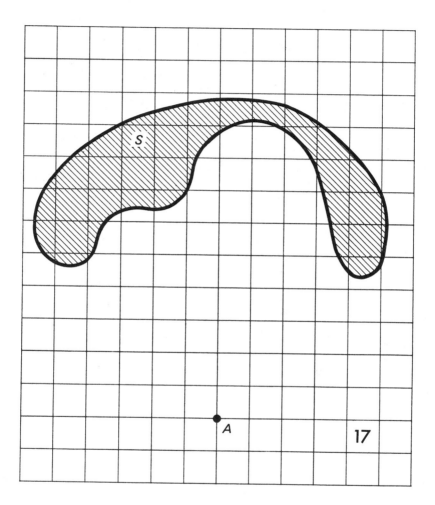

5
TRIANGLES

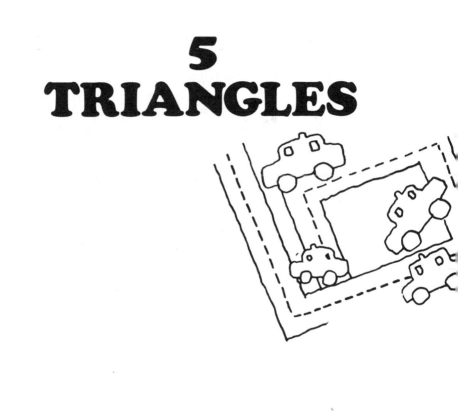

SEVERAL OF the ideas we have met already can be related by introducing a uniform terminology. In Section 1 we took two points, A and B, and investigated all points equidistant from them. Let us refer to $\{P|d_T(P, A) = d_T(P, B)\}$ as the *taxicab midset of A and B*, and to $\{P|d_E(P, A) = d_E(P, B)\}$ as the *Euclidean midset of A and B*. You probably recall that the Euclidean midset of A and B turns out to be the perpendicular bisector of $\overline{AB}$, while the taxicab midset of A and B can have a variety of shapes (See Exercise 11 of Section 1).

In Section 4 we took a point F and a line L and investigated all points equidistant from them. In deference to tradition we called $\{P|d_E(P, F) = d_E(P, L)\}$ a Euclidean parabola and we called $\{P|d_T(P, F) = d_T(P, L)\}$ a taxicab parabola; but we could just as well have referred to these figures as the *Euclidean midset of F and L* and the *taxicab midset of F and L*, respectively.

The next natural step is to take two lines, L_1 and L_2, and consider their midsets. The *Euclidean midset of L_1 and L_2* is $\{P|d_E(P, L_1) = d_E(P, L_2)\}$. The *taxicab midset of L_1 and L_2* is $\{P|d_T(P, L_1) = d_T(P, L_2)\}$. Figure 18 suggests how one finds the taxicab midset of two given lines L_1 and L_2. Begin by finding $\{P|d_T(P, L_1) = 2\}$ and $\{P|d_T(P, L_2) = 2\}$ and intersecting them. These are the fine dashed lines in Fig. 18. This gives four points on the taxicab midset of L_1 and L_2. Then find $\{P|d_T(P, L_1) = 4\}$ and $\{P|d_T(P, L_2) = 4\}$ and intersect them. These are the fine dotted lines in Fig. 18. This gives four more points on the taxicab midset of L_1 and L_2. Now join the points you have found on the taxicab midset in the most natural way. These are the heavy dotted lines in Fig. 18. Why is the intersection point of L_1 and L_2 a member of their midset? Using hindsight, was it necessary to plot $\{P|d_T(P, L_1) = 4$ and $d_T(P, L_2) = 4\}$?

The Euclidean midset of L_1 and L_2 is also a pair of lines which intersect at $L_1 \cap L_2$. For most pairs of lines L_1 and L_2, the Euclidean midset of L_1 and L_2 is different from the taxicab midset of L_1 and L_2. Do you recall how to find the Euclidean midset of L_1 and L_2 using only a protractor?

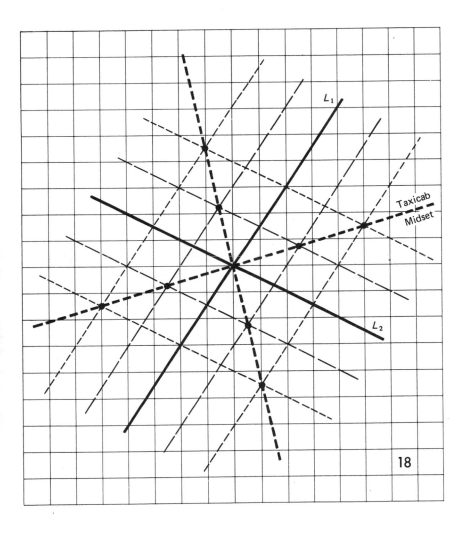

18

exercises

1. On a sheet of graph paper sketch the line L_1 through $(3, 4)$ and $(0, 0)$, and the line L_2 through $(0, 0)$ and $(3, 0)$.

   a) Sketch $\{P | d_T(P, L_1) = 3\}$

   b) Sketch $\{P | d_T(P, L_2) = 3\}$

   c) Sketch the taxicab midset of L_1 and L_2.

   d) Using a different color, sketch the Euclidean midset of L_1 and L_2.

2. On a sheet of graph paper mark the points

$$A = (-3, 0), \qquad B = (-5, 4), \qquad C = (3, 0).$$

   a) Sketch all points P *in the interior* of $\angle BAC$ which are equi-taxi-distant from the sides of the angle.

   b) Using a different color, repeat part (a) for equi-Euclid-distant points.

3. On a sheet of graph paper, mark the points

$$A = (6, -3), \qquad B = (0, 6), \qquad C = (8, 4).$$

   Inscribe a Euclidean circle in $\triangle ABC$. (*Hint.* Its center is a point equidistant from the three sides of the triangle. Locate this point, called the *incenter* of $\triangle ABC$, with the help of a protractor.)

4. On another sheet of graph paper, mark the same points A, B, C as in Exercise 3. Inscribe a taxicab circle in $\triangle ABC$. (*Hint.* First locate its center.) Do you think that, given any triangle at all, it is possible to inscribe a taxi circle in it?

5. For the 3 points $A = (4, 0)$, $B = (1, 8)$, $C = (-6, -2)$, locate 4 Euclidean circles each one tangent to all three of the lines: $\overleftrightarrow{AB}, \overleftrightarrow{AC}, \overleftrightarrow{BC}$.

6. Repeat Exercise 5 for taxicab circles.

7. Using the same points A, B, C as in Exercise 3, *circumscribe* a Euclidean circle about $\triangle ABC$. (*Hint.* Its center is a point equidistant from the three vertices of the triangle. Locate this point, called the *circumcenter* of $\triangle ABC$, with the help of a compass.)

8. Using the same points A, B, C as in Exercise 3, circumscribe a taxicab circle about $\triangle ABC$. (*Hint.* First locate its center.)

9. Circumscribe a taxicab circle about $\triangle ABC$ where $A = (4, 0)$, $B = (1, 8)$, $C = (-6, -2)$. Do you think that, given any triangle at all, it is possible to circumscribe a taxicab circle about it?

10. Try circumscribing a taxicab circle about the triangle with vertices $A = (-3, 0)$, $B = (0, 1)$, $C = (5, 5)$.

11. Repeat Exercise 10 for $A = (-6, 0)$, $B = (0, 6)$, $C = (9, -1)$.

12. Alice is still working at $A = (-3, -1)$, Bruno is back with the bakery at $B = (3, 3)$, and their young son Clyde goes to Coolidge Elementary School at $C = (0, -3)$.

    a) Where should they live so that each of the three of them has the same distance to walk to work or school?

    b) Where should they live if they have decided that Clyde should have the shortest walk, Bruno the second shortest, and Alice the longest?

6
FURTHER APPLICATIONS TO URBAN GEOGRAPHY

TAXICAB GEOMETRY is a better mathematical model than Euclidean geometry for solving problems in urban geography. In some ways, taxicab geometry is a much simpler geometry, too. Consider this pair of problems.

Problem 1—Given three points $A = (2, 4)$, $B = (7, -1)$, and $C = (-3, 1)$. Find a point P for which

$$d_E(P, A) + d_E(P, B) + d_E(P, C)$$

is as small as possible.

Problem 2—Given the same three points A, B, and C, find P for which

$$d_T(P, A) + d_T(P, B) + d_T(P, C)$$

is as small as possible

The first problem is difficult and we shall not attempt to solve it.* The second problem, however, is remarkably simple.

In Fig. 19 we have drawn a horizontal line (solid) through C and a vertical line (dotted) through A. Suppose you begin on the solid line, say at Q, and you walk one unit up, to Q'. In doing this, you decrease your distance from A by one unit, but you increase your distance from B and from C by one unit each. Thus you increase the sum of your distances from A, B, and C:

$$d_T(Q', A) + d_T(Q', B) + d_T(Q', C) > d_T(Q, A) + d_T(Q, B) + d_T(Q, C).$$

In fact, if you begin anywhere on the solid line and move vertically off it, either up or down, you will increase the sum of your distances from A, B, and C. This is because you will always be moving away from more points than you are moving toward.

* For a solution, see pages 21–22 of *Introduction to Geometry* by H. S. M. Coxeter (Wiley, 1961).

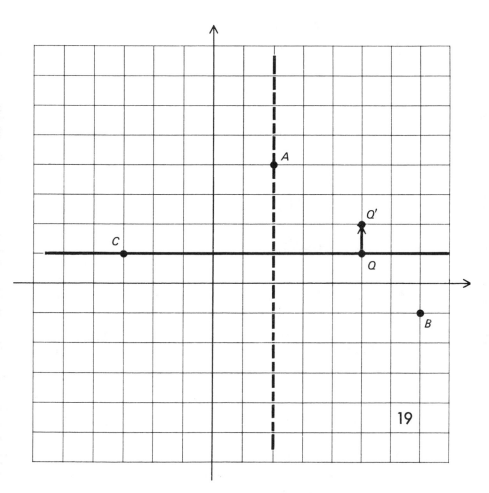

Similarly, any horizontal move off the dotted line will increase the sum of your distances from A, B, and C.

Thus, if you begin at the point of intersection of the solid and dotted lines, a move in *any* direction will increase the sum of the distances from A, B, and C. It follows that

$$d_T(P, A) + d_T(P, B) + d_T(P, C)$$

is minimized when P is the point of intersection of the solid and dotted lines. We have solved Problem 2.

A similar, but somewhat more difficult problem, is the following: Given four points $A = (-1, 4)$, $B = (3, 1)$, $C = (1, -1)$, and $D = (-4, 1)$, find all points P for which

$$d_T(P, A) + d_T(P, B) + d_T(P, C) + d_T(P, D)$$

is as small as possible.

We begin by plotting the points and drawing a horizontal line through D and B. (See Fig. 20.) As before, any vertical movement from this line will increase the sum of the distances to A, B, C, and D. Now, instead of drawing a vertical line, we draw an entire vertical strip having A and C on its edges. Observe that a horizontal movement within this strip, say from $(0, 2)$ to $(1, 2)$, leaves the sum of the distances to A, B, C, and D unchanged. (You get closer to two points, B and C, but farther, by the same amount, from the other two.) A horizontal movement away from this strip, however, say from $(1, 3)$ to $(2, 3)$, increases the sum of the distances. (You are moving away from more points than you are moving toward.)

Thus, all of the points P belonging to the intersection of the horizontal line and the vertical strip make

$$d_T(P, A) + d_T(P, B) + d_T(P, C) + d_T(P, D)$$

as small as possible.

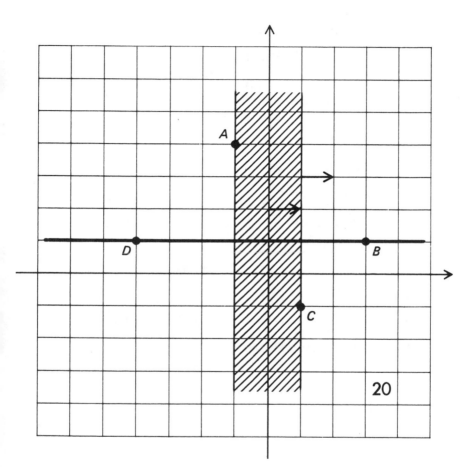

exercises

1. Given a set of points $A, B, C, \ldots$, the point or points P at which

$$d_T(P, A) + d_T(P, B) + d_T(P, C) + \cdots$$

is as small as possible will be called the *minimizing region* of the set. Locate the minimizing region for each of the following sets of points.

a) $A = (-3, 4)$, $B = (4, 3)$, $C = (0, -2)$

b) $A = (-6, 0)$, $B = (2, 4)$, $C = (0, 4)$, $D = (-1, -2)$

c) $A = (-4, 0)$, $B = (-1, 3)$, $C = (3, -1)$, $D = (1, -3)$

d) $A = (-4, 0)$, $B = (-3, 3)$, $C = (0, 2)$, $D = (3, -2)$,
   $E = (-1, -2)$

e) $A = (1, 1)$, $B = (1, 4)$, $C = (6, 1)$

f) $A = (1, 1)$, $B = (3, 1)$, $C = (6, 1)$

g) $A = (0, 0)$, $B = (2, 2)$, $C = (0, 4)$, $D = (-5, 2)$

h) $A = (0, 1)$, $B = (1, 2)$, $C = (2, 0)$, $D = (4, -2)$,
   $E = (1, -1)$

2. Find the minimizing region for each of the following two-point sets.

a) $A = (-2, 3)$, $B = (1, -4)$

b) $A = (1, -3)$, $B = (4, 0)$

c) $A = (2, 1)$, $B = (6, 1)$

d) $A = (1, 1)$, $B = (1, 4)$

Compare your results and your technique with Exercise 8 of Section 1.

3. If a set consists of an odd number of points, what can be said of its minimizing region?

4. Alice, Bruno, and Clyde have to walk to $A = (-3, -1)$, $B = (3, 3)$, and $C = (0, -3)$, respectively. Where should they live so that the sum of the distances they have to walk is a minimum?

5. Clyde's dog Darwin has a job as a night watchman at $D = (-8, -1)$. Where should they look for an apartment so that the sum of the distances the *four* of them have to walk is a minimum?

6. The Chauncey Cement Company wants to locate its plant at a point P for which

$$d_T(P, A) + d_T(P, B) + d_T(P, C)$$

is a minimum, where $A = (-6, 0)$ is a sand quarry, $B = (5, 0)$ is a boat dock, and $C = (-2, 5)$ is a railroad freight yard. Find P.

7. Burger Baron has hamburger stands at $(-5, 5)$, $(-2, 4)$, $(1, 1)$, $(2, 6)$, $(5, -2)$, $(3, -4)$, $(-2, -1)$, and $(-4, -4)$. He wants to build a central supply warehouse so that the sum of the distances to the eight hamburger stands is minimized. Where should the warehouse be located?

8. The Burger Baron stand at $(2, 6)$ does a booming business, requiring twice as many deliveries from central supply as each of the other stands. Bearing this in mind, where should the warehouse be located?

9. The owner of three gasoline stations at $A = (-3, 3)$, $B = (4, 1)$, and $C = (1, -3)$ wants to build a car wash at a point P which will minimize the sum

$$d_T(P, A) + d_T(P, B) + d_T(P, C).$$

a) Where should she build the car wash?

b) What will the (minimum) sum of the distances be?

c) If she decides that it is only necessary to keep the sum less than 15 blocks, where could she build?

*10. Think back to how an ordinary (bifocal) ellipse was defined in taxicab geometry. Now define "the trifocal ellipse with foci A, B, and C and constant 15." Using the points A, B, and C of Exercise 9, draw the family of trifocal ellipses with foci A, B, and C and constants 14, 15, 16, 17, 18, 19, 20, 21, 22, 23, 24, and 25. If the trifocal ellipse with foci A, B, and C and constant 1000 were drawn, and then viewed from a distance, what would it look like?

*11. Investigate the family of four-focal ellipses with foci at $(-4, 2)$, $(0, 7)$, $(4, 0)$, and $(8, -2)$.

In the remainder of this section we alter our taxicab geometry to produce an even more accurate model of urban geography. The gain in realism, however, must be paid for, and the price is increasingly complicated mathematics.

Ideal City's new mass-transit vehicle runs along the line L shown in Fig. 21. Using it decreases certain walking distances. For example, if Bruno is at the point $P = (-3, 6)$ and wants to go to the bakery $B = (3, 3)$, he walks three blocks south to $(-3, 3)$, rides the mass transit to $(3, 1)$, and then walks two blocks north to B. The walking distance from P to B is five. Of course, if Bruno started out at the point $Q = (1, 5)$, he would not use the mass transit. His walking distance to B would be four. We shall name this new distance function "d_M" in honor of the mass-transit system. Thus $d_M(P, B) = 5$ and $d_M(Q, B) = 4$. The formal definition of d_M is as follows:

$d_M(A, B) = $ The smaller of $d_T(A, B)$ and $d_T(A, L) + d_T(B, L)$.

12. a) Compute the mass-transit distance from $(0, 0)$ to $(3, 3)$.

   b) Compute the mass-transit distance from $(-6, 2)$ to $(3, 3)$.

   c) Compute the mass-transit distance from $(6, 4)$ to $(3, 3)$.

   d) Sketch the set of all points at mass-transit distance 4 from the point $(3, 3)$.

   e) What is a reasonable name for the figure in part (d)?

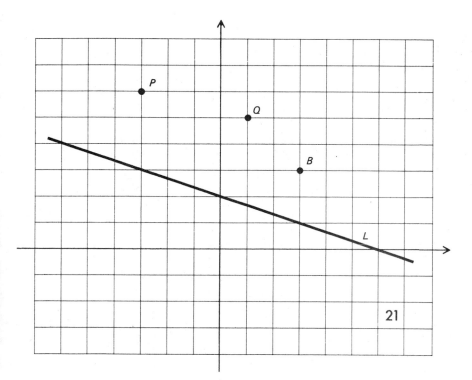

13. Given $B = (3, 3)$. Sketch the mass-transit circle with center B and:

    a) Radius 3

    b) Radius $2\frac{1}{2}$

    c) Radius 2

    d) Radius 1

14. Which of these properties of a *metric** does d_M have?

    a) $d_M(A, B) \geqslant 0$ for all A, B

    b) $d_M(A, B) = 0$ if and only if $A = B$

    c) $d_M(A, B) = d_M(B, A)$

    d) $d_M(A, B) + d_M(B, C) \geqslant d_M(A, C)$

15. Alice works at $A = (-3, -1)$ and Bruno works at $B = (3, 3)$.

    a) Find $d_M(A, B)$

    b) Find $\{P | d_M(P, A) = 3 \text{ and } d_M(P, B) = 3\}$

    c) Find $\{P | d_M(P, A) = 4 \text{ and } d_M(P, B) = 2\}$

    d) Find $\{P | d_M(P, A) = 5 \text{ and } d_M(P, B) = 1\}$

    e) Where could Alice and Bruno live if they want the sum of their mass-transit distances from work to be a minimum?

    f) Where could they live if they want the sum to still be a minimum, but they want Bruno to be "closer" to his job than Alice is to hers?

16. a) Where could Alice and Bruno live if they want only to be equidistant from their jobs?

---

* In the appendix it is noted that d_E and d_T are also metrics, in the sense that they satisfy these same four properties.

16. (*cont'd.*)

   b) Where could they live if it is only important that Alice be closer to her job than Bruno is to his?

17. Draw these mass-transit "ellipses" in different colors on a single sheet of graph paper. $A = (-3, -1)$ and $B = (3, 3)$.

   a) $\{P|d_M(P, A) + d_M(P, B) = 8\}$

   b) $\{P|d_M(P, A) + d_M(P, B) = 7\}$

   c) $\{P|d_M(P, A) + d_M(P, B) = 6\}$

18. Where could Alice and Bruno live if they want the sum of their distances from work to be at most 8 blocks?

19. Draw these mass-transit "hyperbolas" in different colors on a single sheet of graph paper. $A = (-3, -1)$ and $B = (3, 3)$.

   a) $\{P| |d_M(P, B) - d_M(P, A)| = 1\}$

   b) $\{P| |d_M(P, B) - d_M(P, A)| = 2\}$

   c) $\{P| |d_M(P, B) - d_M(P, A)| = 3\}$

   d) $\{P| |d_M(P, B) - d_M(P, A)| = 6\}$

   e) $\{P| |d_M(P, B) - d_M(P, A)| = 0\}$

20. Where could Alice and Bruno live if their only condition is that neither person has to walk more than 2 blocks farther to work than the other person?

21. Sketch the set of all points P at mass-transit distance 1 from the horizontal axis.

7
SOME
DIRECTIONS
FOR FURTHER
RESEARCH

WE HAVE investigated only a few of the questions suggested by taxicab geometry. Probably many others occurred to you as you worked through the preceding sections. It is likely that some of your questions have never been answered, or even asked, before. These questions which you formulate yourself provide an ideal basis for truly original research. We encourage you to follow where your questions lead and to report on your discoveries. In mathematics, curiosity is the mother of invention.

Mathematical questions tend to spring from two main sources: the real world, and the abstract world of mathematics. The real world, with its city streets, provided much of the motivation for studying taxicab geometry in the first place. Questions of optimum locations for apartments, factories, phone booths, etc., produced a host of geometrical problems. The real world also suggested modifying our model from taxicab geometry to mass-transit geometry. This generated another whole array of mathematical questions which we only began to investigate. (Quite a few new problems will arise if you try to do for mass-transit geometry everything that we did for taxicab geometry. For example, study the midset of a point and a line, the midset of two lines, the incenter of a triangle, the circumcenter of a triangle, etc.) Posing new "real" problems will lead to more geometrical questions about both taxicab and mass-transit geometry.

Entire new geometries are also suggested by real-world cities. Change the route of the mass transit. Put a kink in it. Put in two mass-transit lines. Put a pond somewhere in the city. Note that cities have a vertical dimension, and study taxicab geometry in space. The list of questions suggested by the real world seems to be endless.

Purely mathematical considerations can also suggest further questions for study. For example, we might observe that the whole field of taxicab geometry was opened up by simply replacing the Euclidean distance function d_E by the taxicab distance function d_T. Recall that if $A = (a_1, a_2)$ and $B = (b_1, b_2)$, then the precise definitions of d_E and d_T were:

$$d_E(A, B) = \sqrt{(a_1 - b_1)^2 + (a_2 - b_2)^2},$$
$$d_T(A, B) = |a_1 - b_1| + |a_2 - b_2|.$$

A natural mathematical question to ask is this: What happens to the geometry if we replace d_E by some other distance function which need not have any connection with the real world?

In the exercises which follow you can begin to answer this question, using the distance function d_L defined by

$$d_L = \text{The larger of } |a_1 - b_1| \text{ and } |a_2 - b_2|.$$

Two other very broad research projects are these.
(1) Define taxi trig functions via wrapping of the unit taxi circle, and investigate their graphs, trig identities, ...
(2) Instead of beginning with a square grid of streets as in Fig. 3, page 13, begin with an equilateral triangle grid and develop an entire "Chinese Checkers Geometry" parallel to Taxicab Geometry. (I am indebted to my son Tom for the name of this geometry and for some interesting discoveries he made about it.)

exercises

1. Plot each pair of points given, and then find $d_L(A, B)$.

   a) $A = (-2, 1)$, $B = (1, 2)$

   b) $A = (-2, 1)$, $B = (-2, -2)$

   c) $A = (-2, 1)$, $B = (1, -2)$

   d) $A = (-2, 1)$, $B = (-5, 4)$

2. Sketch:

   a) The d_L-circle with center $A = (-2, 1)$ and radius 3

   b) The d_L-circle with center $(2, -4)$ and radius 3

   c) The d_L-midset of $(-2, 1)$ and $(2, -4)$

3. Which of these properties of a metric does d_L have?

   a) $d_L(A, B) \geqslant 0$ for all A, B

   b) $d_L(A, B) = 0$ if and only if $A = B$

   c) $d_L(A, B) = d_L(B, A)$

   d) $d_L(A, B) + d_L(B, C) \geqslant d_L(A, C)$

4. Define d_S in terms of the coordinates of $A = (a_1, a_2)$ and $B = (b_1, b_2)$ as follows:

   $$d_S(A, B) = \text{The smaller of } |a_1 - b_1| \quad \text{and} \quad |a_2 - b_2|$$

   Which of the properties (a) through (d) of a metric does d_S have?

5. Define d_K to be the average of d_T and d_L. That is,

   $$d_K(A, B) = \tfrac{1}{2}[d_T(A, B) + d_L(A, B)]$$

   Which of the properties (a) through (d) of a metric does d_K have? (*Hint.* Both d_T and d_L have all four.)

6. Sketch the d_K-circle with center $(0, 0)$ and radius 4.

APPENDIX

taxicab geometry
and euclidean
geometry compared*

appendix

WE HAVE investigated intuitively two geometries: one based on distance as the crow flies, the other based on distance as a taxicab drives. Each geometry can be thought of more formally as a mathematical system consisting of

1. A set $\mathscr{P}$ of points,
2. A collection $\mathscr{L}$ of subsets of $\mathscr{P}$ called lines,
3. An angle measure function m,
4. A distance function—d_E for ordinary geometry, d_T for taxicab geometry,

An informative symbol for the system of ordinary geometry then is $[\mathscr{P}, \mathscr{L}, d_E, m]$, and the corresponding symbol for taxicab geometry is $[\mathscr{P}, \mathscr{L}, d_T, m]$.

The system $[\mathscr{P}, \mathscr{L}, d_E, m]$ is called Euclidean geometry because it enjoys the thirteen properties which constitute a modern-day set of axioms for describing the geometry promulgated by Euclid in antiquity. We shall not *prove* that $[\mathscr{P}, \mathscr{L}, d_E, m]$ has all these properties. Many of the proofs would require advanced, explicit definitions of lines and the angle measure function m, as well as some quite sophisticated arguments. Instead, we shall merely state the properties and assume that $[\mathscr{P}, \mathscr{L}, d_E, m]$ has them. (You should check each one against your intuition by making a rough sketch.) On the basis of this assumption we shall then decide which of these properties hold true for taxicab geometry and which fail. Perhaps surprisingly we shall see that taxicab geometry enjoys all but one of the thirteen fundamental properties.

The first two properties are known as incidence properties.

1 Given any two points, there is exactly one line containing them.

2 Every line contains at least two points; $\mathscr{P}$ contains at least three noncollinear points.

---

* This rather theoretical appendix is most appropriate for students who have already completed an axiomatic course in Euclidean geometry.

appendix

Since these properties concern only $\mathscr{P}$ and $\mathscr{L}$, they are as true of $[\mathscr{P}, \mathscr{L}, d_\mathrm{T}, m]$ as they are of $[\mathscr{P}, \mathscr{L}, d_\mathrm{E}, m]$.

The next four axioms assert that the distance function is "positive definite," "symmetric," satisfies the "triangle inequality," and has the "ruler property." In detail, these four properties for d_E are as follows.

$\boxed{3}$ To each ordered pair of points (A, B), d_E assigns a non-negative number $d_\mathrm{E}(A, B)$. Moreover, $d_\mathrm{E}(A, B) = 0$ if and only if $A = \mathbf{B}$.

$\boxed{4}$ $d_\mathrm{E}(A, B) = d_\mathrm{E}(B, A)$

$\boxed{5}$ $d_\mathrm{E}(A, B) + d_\mathrm{E}(B, C) \geqslant d_\mathrm{E}(A, C)$

$\boxed{6}$ Given any line L there is a one-to-one, onto function f_L from L to $\mathbb{R}$ (the real numbers) such that for all points A, B on L

$$\left| f_\mathrm{L}(A) - f_\mathrm{L}(B) \right| = d_\mathrm{E}(A, B).$$

appendix

exercises [begun]

Use the explicit algebraic definition of d_T,

$$d_T(A, B) = |a_1 - b_1| + |a_2 - b_2|$$

where

$$A = (a_1, a_2), \qquad B = (b_1, b_2),$$

and anything you know about *absolute value* to prove the following,

1. a) $d_T(A, B) \geq 0$.

   b) If $d_T(A, B) = 0$, then $A = B$.

   c) If $A = B$, then $d_T(A, B) = 0$.

2. $d_T(A, B) = d_T(B, A)$.

3. $d_T(A, B) + d_T(B, C) \geq d_T(A, C)$. *Hint.* Use this fact about absolute value: for all numbers x and y,

$$|x + y| \leq |x| + |y|$$

To prove that $[\mathscr{P}, \mathscr{L}, d_T, m]$ has property $\boxed{6}$, we break it down into two cases—Case 1: L is vertical; Case 2: L is not vertical.

4. In the case where L is vertical, define f_L by

$$f_L(x_1, x_2) = x_2 \qquad \text{for all } (x_1, x_2) \in L.$$

a) Check that f_L is a one-to-one, onto function from L to $\mathbb{R}$.

b) Verify that for all points A, B on L

$$|f_L(A) - f_L(B)| = d_T(A, B).$$

*5. In the case where L is not vertical, L has a finite "slope" m. That is, there is a number m such that if $A = (a_1, a_2)$ and $B = (b_1, b_2)$ are any two points of L, then

$$\frac{a_2 - b_2}{a_1 - b_1} = m.$$

70

In this case, define f_L by

$$f_L(x_1, x_2) = (1 + |m|) \cdot x_1 \qquad \text{for all } (x_1, x_2) \in L.$$

a) Check that f_L is a one-to-one, onto function from L to $\mathbb{R}$.

b) Check that, for all points A, B on L

$$|f_L(A) - f_L(B)| = d_T(A, B).$$

Next on the list of thirteen properties is the plane separation property.

$\boxed{7}$ If L is any line, then there exist subsets H_1 and H_2 of $\mathscr{P}$ (called half-planes) such that

   i) H_1 and H_2 are convex;

   ii) $H_1 \cup H_2 = \mathscr{P} - L$ ($\mathscr{P}$ with L removed);

   iii) If $A \in H_1$ and $B \in H_2$, then $\overline{AB} \cap L \neq \varnothing$.

On the surface this property does not seem to involve d_E, so if it is true for $[\mathscr{P}, \mathscr{L}, d_E, m]$, then it should also be true for $[\mathscr{P}, \mathscr{L}, d_T, m]$. On closer inspection, though, we see that it involves the concepts of *segment* and *convexity*, which ultimately rest on a distance function. To check that $[\mathscr{P}, \mathscr{L}, d_T, m]$ has the plane-separation property, then, we need to review the sequence of concepts leading up to segment and convexity.

exercises {continued}

In any geometry with a distance function d, the definition of betweenness is as follows: P is *between* A and B (written $A-P-B$) if and only if A, P, B are distinct points such that

   i) $d(A, P) + d(P, B) = d(A, B)$,

   ii) A, P, B are collinear

If we replace d by d_E, we get a definition of betweenness in Euclidean geometry. If we replace d by d_T, we get a definition of betweenness in taxi geometry.

6. Given $A = (-2, -1)$ and $B = (3, 2)$.

    a) Graph $\{P|d_T(P, A) + d_T(P, B) = d_T(A, B)\}$

    b) On the same sheet of graph paper, graph

    $$\{P|A, P, B \text{ are collinear}\}.$$

    c) On the same sheet of graph paper, graph

    $$\{P|P \text{ is taxi-between } A \text{ and } B\}.$$

7. a) Does condition (ii) in the definition of betweenness have any significance in Euclidean geometry?

    b) Sketch $\{P|P \text{ is Euclid-between } A \text{ and } B\}$, where A and B are as in Exercise 6.

    c) Does betweenness mean the same thing in both taxicab and Euclidean geometry? That is, is it true that P is taxi-between A and B if and only if P is Euclid-between A and B?

8. a) How is a segment $\overline{AB}$ defined in terms of betweenness?

    b) Are segments in taxicab geometry the same as segments in Euclidean geometry?

    c) How is convexity defined in terms of segments?

    d) Does convexity mean the same thing in taxicab geometry as in Euclidean geometry? That is, if a subset H of $\mathscr{P}$ is taxi-convex, is it also Euclid-convex, and vice versa?

9. Assuming that $[\mathscr{P}, \mathscr{L}, d_E, m]$ has the plane-separation property, does $[\mathscr{P}, \mathscr{L}, d_T, m]$ have it too?

10. a) How is a ray $\overrightarrow{AB}$ defined in terms of betweenness?

    b) Is every taxi-ray a Euclid-ray, and vice versa?

    c) How is an angle $\angle ABC$ defined in terms of rays?

    d) Is every taxi-angle a Euclid-angle, and vice versa?

exercises

11. a) If $A \notin \overleftrightarrow{BC}$, what is meant by "the half-plane, A's side of $\overleftrightarrow{BC}$"?

   b) Is A's side of $\overleftrightarrow{BC}$ in taxi geometry the same as A's side of $\overleftrightarrow{BC}$ in Euclidean geometry?

   c) How is the interior of an angle $\angle ABC$ defined in terms of half-planes?

   d) Does "interior of an angle" mean the same thing in both taxicab and Euclidean geometry?

Next on the list of thirteen properties is a block of four angle-measure properties.

8   m assigns to each angle a real number between 0 and 180.

9   Given a ray $\overrightarrow{AB}$ on the edge of a half-plane H and given a real number r between 0 and 180, there is exactly one ray $\overrightarrow{AP}$ such that $P \in H$ and $m\angle PAB = r$.

10   If D is in the interior of $\angle ABC$, then

$$m\angle ABD + m\angle DBC = m\angle ABC.$$

11   If B is between A and C and $D \notin \overleftrightarrow{AC}$, then

$$m\angle ABD + m\angle DBC = 180.$$

These properties must be true of $[\mathscr{P}, \mathscr{L}, d_T, m]$ if they are true of $[\mathscr{P}, \mathscr{L}, d_E, m]$, since both geometries use the same angle measure function m and since "angle," "ray," "half-plane," "interior of an angle," and "between" have the same meaning in both geometries.

The next property of $[\mathscr{P}, \mathscr{L}, d_E, m]$ is the "side-angle-side" property.

12   Given a one-to-one correspondence between the vertex sets of two triangles. If two sides and the included angle of the first triangle are congruent to the corresponding parts of the second triangle, then the correspondence is a congruence.

This is the one basic property of $[\mathscr{P}, \mathscr{L}, d_E, m]$ that $[\mathscr{P}, \mathscr{L}, d_T, m]$ does *not* have. That is why $[\mathscr{P}, \mathscr{L}, d_T, m]$ is called a *non-Euclidean*

appendix

geometry. We will investigate this side-angle-side property more fully in the exercises which follow, but first we should complete the list of thirteen properties. The last property of $[\mathscr{P}, \mathscr{L}, d_E, m]$ is the famous parallel property.

13 Given a point P off a line L, there is exactly one line through P parallel to L.

This property must certainly be true of $[\mathscr{P}, \mathscr{L}, d_T, m]$ as well, since it has only to do with $\mathscr{P}$ and $\mathscr{L}$.

exercises (concluded)

12. Two triangles, $\triangle ABC$ and $\triangle A'B'C'$ are shown in Fig. 22.

   a) Does $d_T(A, B) = d_T(A', B')$?

   b) Does $d_T(A, C) = d_T(A', C')$?

   c) Does $m \angle BAC = m \angle B'A'C'$?

   d) Under the correspondence

   $$A \leftrightarrow A', B \leftrightarrow B', C \leftrightarrow C',$$

   do $\triangle ABC$ and $\triangle A'B'C'$ have side-angle-side of one congruent respectively to the corresponding side-angle-side of the other?

   e) Is $\triangle ABC \cong \triangle A'B'C'$? Why or why not?

13. When "SAS" fails, it pulls down with it a lot of other familiar congruence conditions. In taxicab geometry do the following.

   a) Exhibit a pair of incongruent triangles for which "ASA" holds.

   b) Exhibit a pair of incongruent triangles for which "SAA" holds.

   c) Exhibit a pair of incongruent triangles for which "SSS" holds.

74

exercises

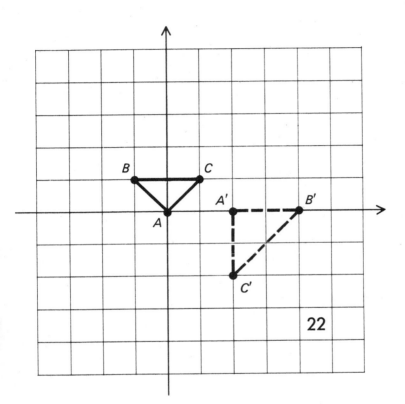

14. Still in taxicab geometry:

    a) Exhibit an isosceles triangle whose base angles are incongruent.

    b) Exhibit an equilateral right triangle.

    c) Exhibit a triangle with two congruent angles which is not isosceles.

SELECTED
ANSWERS

selected answers

Section 1

Page 6, Ex. 1	(a) $6, \sqrt{20}$
	(e) 6, 6
Page 6, Ex. 2	(a) No. See Exercise 1, parts (a) and (e).
Page 6, Ex. 2	(d) Hint. Show, by squaring both sides, that if x and y are nonnegative numbers, then $x + y \geqslant \sqrt{x^2 + y^2}$.
Page 7, Ex. 4	(b) See Fig. 23
	(d) Taxi circle with center A and radius 3
Page 7, Ex. 6	(b) The line segment joining $(-2, 2)$ to $(1, -1)$
Page 8, Ex. 8	(a) The shaded rectangle in Fig. 24
	(c) The segment in Fig. 24
Page 8, Ex. 10	(d) The perpendicular bisector of $\overline{AB}$
Page 8, Ex. 11	(a) See Fig. 25
Page 9, Ex. 13	To prove analytically that this figure is a (Euclidean) circle reduce the distance equation $$(x + 3)^2 + y^2 = 4[(x - 1)^2 + (y - 2)^2]$$ to the form $$(x - a)^2 + (y - b)^2 = r^2$$ (In view of this result it is rather surprising that the figure in Exercise 12 is *not* a taxi circle.)

Section 2

Page 14, Ex. 3	Inside the rectangle with opposite vertices A and B but on or to the left of the line segment joining $(-2, 3)$ to $(2, -1)$.
Page 16, Ex. 9	Draw taxicab circles of radius 3 with centers at the fountains. Then decide.
Page 16, Ex. 10	Nine
Page 16, Ex. 12	Hint. Begin by ignoring Harding and finding the Fillmore–Grant boundary line.

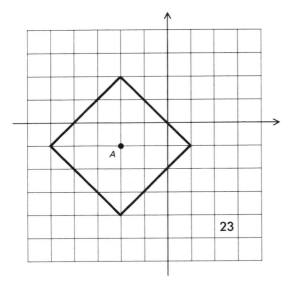

23

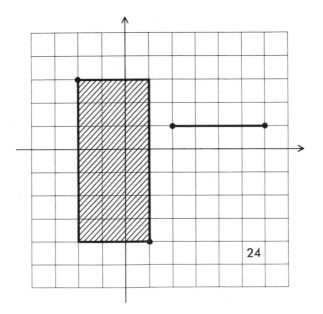

24

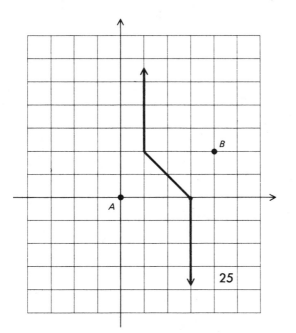

25

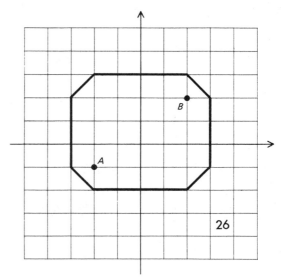

26

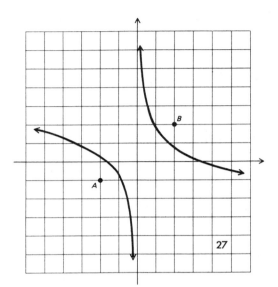

27

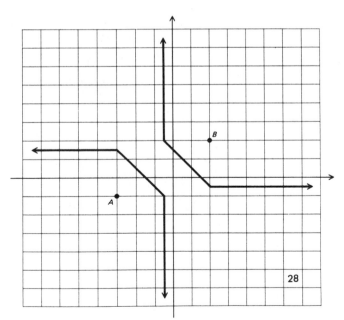

28

Section 3
Page 26, Ex. 1   (f) No points of intersection

Page 26, Ex. 3   (a) The set in Exercise 2(e) is empty because of the triangle inequality:
$$d_E(P, A) + d_E(P, B) \geqslant d_E(A, B) = 5$$

Page 27, Ex. 4   See Fig. 26

Page 27, Ex. 8   Caution. Although it doesn't say so, the noise ordinance must mean a *Euclidean* distance of three blocks. Why?

Page 28, Ex. 9   Fig. 27

Page 29, Ex. 11   See Fig. 28

Section 4
Page 34, Ex. 1   (e) 13

Page 34, Ex. 2   (b) 3

Page 34, Ex. 3   (b) 4

Page 37, Ex. 7   (a) The heavy lines in Fig. 29
                    (c) The dotted lines in Fig. 29

Page 39, Ex. 13 (b) See Fig. 30

Page 42, Ex. 16 (b) Very nearly 8

Section 5
Page 48, Ex. 1   (c) The heavy lines in Fig. 31
                    (d) The dotted lines in Fig. 31

Page 48, Ex. 2   (a) The ray from A through $(0, 6)$
                    (b) The bisector of $\angle BAC$

Page 48, Ex. 5   Hint. These three lines form six pairs of vertical angles. Draw and extend the six angle bisectors and look for points where three of them concur.

Section 6
Page 56, Ex. 1   (a) $(0, 3)$

Page 57, Ex. 9   (b) 13

Page 58, Ex. 12 (b) 4

Page 58, Ex. 12   (d) The union of these 6 figures: line through $(0, 0)$ and $(3, -1)$, ray from $(0, 4)$ through $(-3, 5)$, ray from $(6, 2)$ through $(12, 0)$, segment joining $(0, 4)$ to $(3, 7)$, segment joining $(3, 7)$ to $(7, 3)$, segment joining $(7, 3)$ to $(6, 2)$.

Section 7
Page 66, Ex. 1   (a) 3

Page 66, Ex. 2   (a) A Euclidean square with opposite vertices $(1, -2)$ and $(-5, 4)$

Appendix
Page 70, Ex. 1   (c) $A = B$

$$\Rightarrow (a_1, a_2) = (b_1, b_2)$$
$$\Rightarrow a_1 = b_1 \text{ and } a_2 = b_2$$
$$\Rightarrow a_1 - b_1 = 0 \text{ and } a_2 - b_2 = 0$$
$$\Rightarrow |a_1 - b_1| = 0 \text{ and } |a_2 - b_2| = 0$$
$$\Rightarrow |a_1 - b_1| + |a_2 - b_2| = 0$$
$$\Rightarrow d_T(A, B) = 0$$

Page 70, Ex. 5   (b) $d_T(A, B)$

$$= |a_1 - b_1| + |a_2 - b_2|$$
$$= |a_1 - b_1| + |m \cdot (a_1 - b_1)|$$
$$= |a_1 - b_1| + |m| \cdot |a_1 - b_1|$$
$$= (1 + |m|) \cdot |a_1 - b_1|;$$
$$f_L(A) - f_L(B)$$
$$= |(1 + |m|) \cdot a_1 - (1 + |m|) \cdot b_1|$$
$$= |(1 + |m|) \cdot (a_1 - b_1)|$$
$$= |1 + |m|| \cdot |a_1 - b_1|$$
$$= (1 + |m|) \cdot |a_1 - b_1|$$

Page 72, Ex. 7   (c) Yes

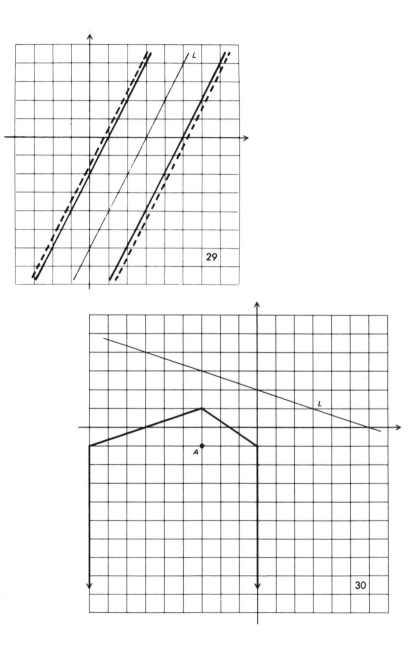

29

30

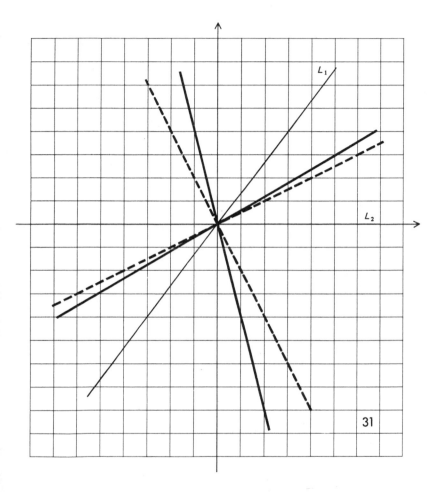

selected answers

Page 72, Ex. 8   (a) $\overline{AB}$ = the set consisting of A, B, and all
                 points between A and B
                 $$= \{A\} \cup \{B\} \cup \{P|A-P-B\}$$
                 (c) A set S is convex means that whenever A
                 $\in S$
                 and $B \in S$ then $\overline{AB} \subset S$, too.

Page 72, Ex. 10   (a) AB = $\{A\}$ $\cup$ $\{P|A-P-B\}$ $\cup$ $\{B\}$ $\cup$
                 $\{Q|\overrightarrow{A-B-Q}\}$
                 (d) Yes

Page 73, Ex. 11   (d) Yes

Page 74, Ex. 12   (e) No. The corresponding parts $\overline{BC}$ and $\overline{B'C'}$
                 are not congruent.

INDEX

index